FOCUS ON

ELEMENTARY

CHEMISTRY

Teacher's Manual
3rd Edition

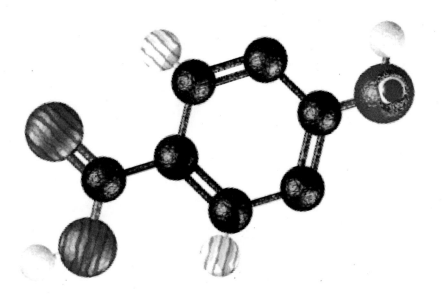

Rebecca W. Keller, PhD

Real Science-4-Kids

Illustrations: Janet Moneymaker

Focus On Elementary Chemistry Teacher's Manual—3rd Edition
ISBN 978-1-941181-38-6

Published by Gravitas Publications Inc.
www.gravitaspublications.com
www.realscience4kids.com

A Note From the Author

This curriculum is designed to provide an introduction to chemistry for students in the elementary level grades. *Focus On Elementary Chemistry—3rd Edition* is intended to be used as the first step in developing a framework for the study of real scientific concepts and terminology in chemistry. This *Teacher's Manual* will help you guide students through the series of experiments in the *Laboratory Notebook*. These experiments will help the students develop the skills needed for the first step in the scientific method — making good observations.

There are several sections in each chapter. The section called *Observe It* helps the students explore how to make good observations. The *Think About It* section provides questions for the students to think about and use to make further observations. In every chapter there is a *What Did You Discover?* section that gives the students an opportunity to summarize the observations they have made. A section called *Why?* provides a short explanation of what students may or may not have observed. And finally, in each chapter there is a section called *Just For Fun* that contains an additional activity.

The experiments take up to 1 hour. The materials needed for each experiment are listed on the next page and also at the beginning of each experiment.

Enjoy!

Rebecca W. Keller, PhD

Materials at a Glance

Experiment 1	Experiment 3	Experiment 5	Experiment 6	Experiment 7

watercolor paints
water in a container
paintbrush
several pieces of paper to paint on
scissors
tape
	magnifying glass
	household items
	   such as:
	     cotton balls
	     rubber bands
	     pencil
	several food items
	   such as:
	     crackers
	     cheese
	     marshmallow
	     beans
	     color-coated candy
		4 or more clear plastic
		   cups or glasses
		marking pen
		measuring cup
		measuring spoons
		the following food
		   items:
		     lemon juice - 180 ml
		     vinegar - 180 ml
		     milk - 180 m
		     baking soda - 90 ml
		     water - 180 ml
		Just For Fun section:
		baking soda
		vinegar
		sugar
		Or
		2 or more food items
		   chosen by student
			clear plastic cups, 12+
			measuring cup
			measuring spoons
			marking pen
			1 head of red cabbage
			knife
			cooking pot, large
			distilled water, 1.25-
			   1.75 liters
			white grape juice, 60 ml
			milk, 60 ml
			lemon juice, 60 ml
			grapefruit juice, 60 ml
			mineral water, 60 ml
			antacid tablets—3 extra-
			   strength unflavored
			   white Tums
			baking soda, 5 ml
			other substances to test
			
			Optional
			small plastic bag
			wooden mallet or
			   other hard object
				18 or more clear plastic
				   cups
				measuring cup
				measuring spoons
				marking pen
				leftover red cabbage
				   juice from
				   Experiment 6 or one
				   head of red cabbage
				food items, approx.
				   300 ml each:
				     vinegar, lemon
				     juice, mineral water,
				     distilled water
				baking soda (25 ml or
				   more)
				antacid tablets, 5 or
				   more (try Tums
				   plain, white, extra
				   strength)
				substances of students'
				   choice

Experiment 2	Experiment 4

salt, 15 ml
water, 237 ml
brick of modeling clay
   (1 or 2)
sugar
	Legos
	small marshmallows,
	   1 pkg
	large marshmallows,
	   1 pkg
	toothpicks

Experiment 8	Experiment 9	Experiment 10	Experiment 11	Experiment 12

15 or more clear plastic
   cups
measuring cup
measuring spoons
spoon for mixing
liquid soap
marking pen
the following food
   items (approx. 60 ml
   each):
     water
     milk
     juice
     vegetable oil
     melted butter
	several glasses or
	   plastic cups
	measuring cup
	3 bags (small paper or
	   plastic)
	several small rocks
	   (5-10)
	Legos (handful)
	sand (2 handfuls)
	sugar (handful)
	salt (2 handfuls)
	water
	food coloring, several
	   colors
	1-2 white coffee filters
	white paper, several
	   sheets
	scissors
	several pencils
	tape
		Elmer's white glue,
		   approx. 30-60 ml
		liquid laundry starch,
		   approx. 30-60 ml (or
		   cornstarch, borax,
		   and water mixture)
		measuring cup
		2 plastic cups
		30 metal paper clips
		Elmer's white glue,
		   approx. 30-60 ml
		Just For Fun section:
		   non-toxic glue
		   such as blue glue,
		   clear glue, wood
		   glue, glitter glue, or
		   paste glue, approx.
		   30-60 ml
		
		Optional
		food coloring
			the following foods:
			   marshmallows (2-3)
			   ripe banana
			   green banana
			   several pretzels or
			   salty crackers
			   raw potato
			   cooked potato
			   other food items
			blindfold
				flour, 2 liters
				1 pkg. active dry yeast
				sugar, 30 ml
				vegetable oil
				salt, 5 ml
				soft butter, 120 ml
				double-acting baking
				   powder, 15 ml
				milk, 360 ml
				measuring cups
				measuring spoons
				4 mixing bowls
				mixing spoon
				floured bread board
				2 cookie sheets
				marking pen
				refrigerator
				oven
				timer
				
				Optional
				rolling pin
				biscuit cutter
				2 bread pans

Materials: Quantities Needed for All Experiments

Equipment	Materials	Foods

blindfold
bowls, mixing, 4
bread board, floured
cookie sheets, 2
cup, measuring
cups or glasses, 12-50 clear plastic
knife
Legos
magnifying glass
oven
pot, large, cooking
refrigerator
scissors
spoon for mixing
spoons, measuring
timer
 	
Optional
biscuit cutter
bread pans, 2
mallet, wooden, or other hard object
rolling pin
	bags (small paper or plastic), 3
	clay, modeling, 1 or 2 bricks
	coffee filters, 1-2 white
	food coloring, several colors
	glue, Elmer's white, approx. 30-60 ml
	glue, non-toxic, such as blue glue, clear glue, wood glue, glitter glue, or paste glue, approx. 30-60 ml
	
	household items such as:
	   cotton balls
	   rubber bands
	   pencil
	paintbrush
	paints, watercolor
	paper, several pieces to paint on
	paper, white, several sheets
	paperclips, 30 metal
	pen, marking
	pencils, several
	rocks, small (5-10)
	sand (2 handfuls)
	soap, liquid
	starch, liquid laundry, approx. 30-60 ml (or cornstarch, borax, and water mixture)
	tape
	toothpicks
	water
	water, distilled
	
	Optional
	bag, 1 plastic, small
		antacid tablets—8 or more extra-strength unflavored white Tums
		baking powder, double-acting
		baking soda
		bananas, 1 ripe, 1 green
		butter
		cabbage, red, 1-2 heads
		flour, 2 liters
		food items such as:
		   beans
		   candy, color-coated (e.g., M&Ms)
		   cheese
		   crackers
		   marshmallow
		juice, grapefruit
		juice, lemon
		juice, white grape
		marshmallows, large, 1 pkg
		marshmallows, small, 1 pkg
		milk
		potato, 1 raw, 1 cooked
		pretzels or salty crackers, several
		salt
		sugar
		vegetable oil
		vinegar
		water, mineral
		yeast, 1 pkg. active dry
		
		Optional
		food items chosen by student

Contents

Experiment 1

Chemistry Every Day

Materials Needed

- watercolor paints
- water in a container
- paintbrush
- several pieces of paper to paint on
- scissors
- tape

Objectives

In this experiment students explore how chemistry is involved in activities they perform daily.

The objectives of this lesson are for the students to:

- Observe their activities.
- Make the connection that the activities they perform involve some aspect of chemistry.

Experiment

I. Think About It

Read this section of the *Laboratory Notebook* with your students.

❶-❷ Have the students think about and make a list of the activities they perform in a day. Some suggestions are:

- *Brushing their teeth.*
- *Shampooing their hair.*
- *Eating a cooked egg.*
- *Washing something with soap.*
- *Using a car or other motorized vehicle for transportation.*

❸ Guide the students in their exploration of whether or not these activities involve chemistry.

There are no right answers for these questions. Just allow the students to explore their own ideas.

II. Observe It

Read this section of the *Laboratory Notebook* with your students.

Have your students observe and make a list of everything they do during the course of one day. Have them be as specific as possible.

Have them observe when they are using products such as toothpaste, shampoo, soap, other cleaning products, or medications.

Have them observe the use of any tools or machines. Are they riding a bike, riding in a car, or using an electric powered tool?

III. What Did You Discover?

Read the questions with your students.

❶-❹ Have the students answer the questions. These can be answered orally or in writing. Again, there are no right answers, and their answers will depend on what they actually observed.

IV. Why?

Read this section of the *Laboratory Notebook* with your students.

Discuss any questions that might come up.

V. Just For Fun

Read this section of the *Laboratory Notebook* with your students.

❶-❹ Help your students use watercolors to observe how colors mix. Have them notice how colors change and which colors mixed together will make black.

❺ If the students are interested, they may enjoy experimenting with mixing different colors of their own choice to see what happens.

❻ Have the students cut out some examples of their paint mixtures and tape them in their *Laboratory Notebook* in the space provided. A hair dryer can be used to speed drying time.

Experiment 2

The Clay Crucible

Materials Needed

- salt, 15 ml (1 Tbsp.)
- water, 237 ml (1 cup)
- brick of modeling clay (1 or 2)
- sugar

Objectives

In this experiment, students will explore how to make and use a basic chemistry tool.

The objectives of this lesson are for students to:

- Create a suitable tool to use in experiments.
- Understand how tools help solve problems.

Experiment

Introduction

Read this section of the *Laboratory Notebook* with your students.

Discuss any questions students may have.

I. Think About It

Read this section of the *Laboratory Notebook* with your students.

Have the students think about how chemistry tools help chemists do specific experiments. Discuss how beakers help chemists measure the volume of liquids and how a balance or scale helps chemists measure the weight of solids.

Explore open inquiry with questions such as the following:

- *Can you hold liquid water in your hand? Why or why not?*
- *Can you hold pebbles in your hand? Why or why not?*
- *Can you use a scale to measure liquid water by pouring water on the scale? Why or why not?*
- *Do you think you could use a beaker to measure large pebbles? Why or why not?*

Have the students think about how these two different tools (beakers and scales) are used by chemists to measure different types of chemicals—liquid chemicals and solid or dry chemicals.

Have the students think of some ways that might be used to separate the salt out of saltwater, and then have them record their ideas in the box provided. Let them use their imagination. There are no right answers to this section.

II. Observe It

Read this section of the *Laboratory Notebook* with your students.

A crucible is a dish-shaped or cup-shaped tool used in chemistry labs. Have your students create a small crucible from the clay. Guide them to notice that they can create a shallow crucible or a deep one. Discuss with them how a shallow crucible will allow the water to evaporate more quickly and a deep, narrow crucible will slow the evaporation down. Have them create a suitable crucible.

Have the students mix the saltwater, making sure the salt is completely dissolved. Have them pour the saltwater into the crucible and then observe what happens over the course of several hours or days. In hot, dry climates, the water will evaporate more quickly, and in cold, humid environments, the water will evaporate more slowly. Have them record what they observe, noting the date and the time of the observation. They may make their observations on the same day or on different days.

III. What Did You Discover?

Read this section of the *Laboratory Notebook* with your students.

❶-❹ Have the students answer the questions. These can be answered orally or in writing. There are no right answers and their answers will depend on what they actually observed.

IV. Why?

Read this section of the *Laboratory Notebook* with your students.

Discuss any questions that might come up.

V. Just For Fun

Have the students repeat the experiment with a sugar/water mixture or a salt/sugar/water mixture. If there is enough clay left over, they might like to make another crucible, trying both mixtures and comparing the results. By adding the sugar (and salt, if used) a little at a time, they can experiment to see how much can be added before it will no longer dissolve.

They are asked to first record what they think will happen and then what actually happens during the experiment.

What Is It Made Of?

Materials Needed

- magnifying glass
- household items such as:
    cotton balls
    rubber bands
    pencil
- several food items such as:
    crackers
    cheese
    marshmallow
    beans
    color-coated candy
      (such as M&Ms)

Objectives

In this experiment students will learn how to make good observations.

The objectives of this lesson are:

- To have younger students make careful observations by noticing details.
- To help students develop a vocabulary to describe their observations.

Experiment

I. Think About It

Read this section of the *Laboratory Notebook* with your students.

Here the students will think about and describe the features of objects such as a cracker, a piece of cheese, a piece of candy. **Without allowing the students to look at the object**, name the object and have the students describe it, using both words and pictures. (They will observe the actual object in the following section.)

Direct their inquiry with questions. For example:

- *What color is a cracker?*
- *Is a cracker hard (like a plastic toy) or soft (like a feather)?*
- *Is a cracker large (like an elephant) or small (like a mouse)?*
- *Is a cracker smooth (like a marble) or rough (like sandpaper)?*
- *If you break a cracker, does it look the same on the inside as the outside?*

Using a cracker as an example, the students' answers may look something like the following example.

(Answers may vary.)

Write down the name of an object. Using words and drawings, describe any features you think it has.

cracker

brown	round	scratchy	crumbly

II. Observe It

Read this section of the *Laboratory Notebook* with your students.

Have the students look carefully at the object you've provided for them. Using the magnifying glass, have them examine the object and make careful observations about it. Ask them if the object looks different from what they thought it would.

Direct their investigation with questions such as the following, again using the cracker as an example.

- *Is the cracker as large (or as small) as you thought it would be?*

- *Is the cracker smooth or rough?*

- *What color is the cracker? Is it exactly brown (or white)? Does it have other colors in it?*

- *What happens to the cracker if you break it in half? Is it the same on the inside as on the outside?*

- *What does the cracker look like under the magnifying glass? Can you describe what you see?*

Often students discover that they have not seen or thought about some detail of a familiar object. For example, sometimes crackers have holes on the top. This may be a detail they have never noticed. Or there may be stripes or speckles in the cracker that they haven't observed before. Also, they can observe that some objects are the same on the inside as they are on the outside, like cheese and cotton balls, but other things are not the same on the inside and the outside, like color-coated candy or beans.

For this part of the experiment (using the cracker as an example), the students' answers may look something like this:

(Answers may vary.)

Write down the name of the object you thought about. Describe what you actually see, using words and drawings.

cracker

white and brown
speckles   holes on top   rough inside   seeds

Have the students repeat this exercise with two or three more objects, first describing the item in the *Think About It* section without looking at the object and then using the *Observe It* section to record what they actually see. They can observe as many items as they want to, describing the item first without looking at it, and then carefully observing the item with a magnifying glass. They may want to choose their own items to observe.

III. What Did You Discover?

Read the questions with your students.

❶-❻ The questions in this section of the *Laboratory Notebook* can be answered verbally or in writing, depending on the writing ability of the students. Help the students think about their observations as they answer these questions. Have the students compare their list of descriptions of each object before and after they looked at it, and help them notice where an observation was the same as what they expected to see and where an observation differed from their expectations.

IV. Why?

Read this section of the *Laboratory Notebook* with your students.

Have a discussion with the students about why their observations may have been different from what they thought they would see.

V. Just For Fun

Have the students choose a person to compare to themselves. They may want to choose someone who is in the room so they can look at that person as they do this part of the experiment.

Guide the students in making observations about what is similar and different between themselves and the other person. Encourage the students to notice details, and have them record their observations. The students' answers will be based on what they actually observe, and there are no right answers to this experiment.

Experiment 4

Follow the Rules!

Materials Needed

- Legos
- small marshmallows, 1 pkg
- large marshmallows, 1 pkg
- toothpicks

Objectives

In this experiment, the students will explore the way the building blocks of matter (atoms) fit together to make molecules.

The objectives of this lesson are to help students understand that:

- Matter is made of smaller units called atoms that combine to form molecules.
- Atoms follow specific rules when forming molecules.

Experiment

Before the students begin the experiment, go through the following introduction with them.

① Place Legos on a table and have the students look at them. Help the students observe that each Lego has a certain number of holes on one side and the same number of pegs on the other side.

② Help the students understand that each Lego has only a certain number of pegs and a certain number of holes. Therefore, only a certain number of structures can be built by attaching more Legos to a foundational Lego block's pegs and holes.

③ Direct the students' inquiry with questions such as the following:

- *How many Legos can you attach to the 2-pegged Lego?*
- *How many Legos can you attach to the 4-pegged Lego?*
- *How many Legos can you attach to the 16-pegged Lego?*

④ Compare the Legos to atoms. Help the students understand that atoms are like Legos in that they can only hook to other atoms in certain ways.

I. Think About It

Read this section of the *Laboratory Notebook* with your students.

❶-❺ Have the students answer the questions in this section. There are no right answers. The objective is to have the students begin to think about molecules having specific structures.

II. Observe It

Read this section of the *Laboratory Notebook* with your students.

Save all the marshmallow molecules the students make. They will be used later in the experiment. Keep the molecules made without rules separate from those made with rules.

❶ Have the students make "molecules" with the marshmallows and toothpicks. First, they will make as many different molecules as they can without using any rules. Have them make several and then draw one of them. An example is shown.

(Answers and drawings may vary.)

❶ **Make as many different "molecules" as you can with the marshmallows and toothpicks.**

   How many can you make? ___10___

   Can you draw one?

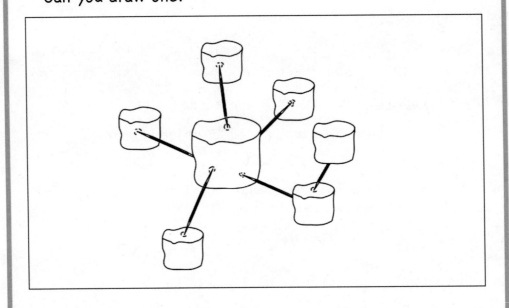

❷ Next, have the students apply a rule to a big marshmallow. The rule is: *Only three toothpicks can be put into a big marshmallow.* They will put a small marshmallow on the other end of each toothpick. Have them make as many shapes as they can think of. Their answer may look something like this:

(Answers and drawings may vary.)

❷ This time you can put only three toothpicks into a big marshmallow. Following this rule, make as many different molecules as you can. A small marshmallow will go on the other end of each toothpick.

How many can you make? ___5___

Can you draw one?

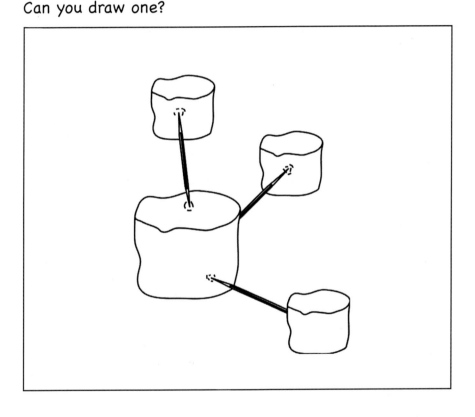

❸ Have the students apply a new rule to a big marshmallow. The new rule is: ***Only two toothpicks can be put into a big marshmallow.*** They will put a small marshmallow on the other end of each toothpick. Their answer may look something like this:

(Answers and drawings may vary.)

❸ Now you can put only two toothpicks into a big marshmallow. Following this rule, make as many different molecules as you can. A small marshmallow will go on the other end of each toothpick.

How many can you make?   3

Can you draw one?

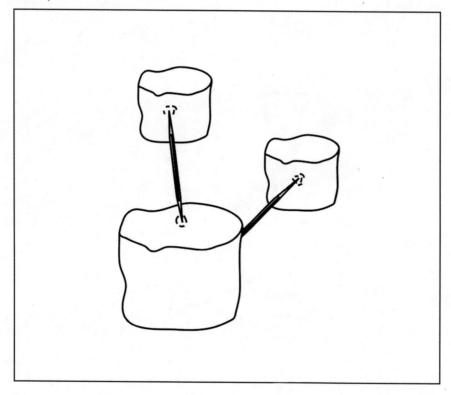

❹ Have the students apply the next rule to a big marshmallow. The rule now is: ***Only one toothpick can be put into a big marshmallow.*** They are to put a small marshmallow on the other end of the toothpick. Their answer may look something like this:

(Answers and drawings may vary.)

❹ Now you can put only one toothpick into a big marshmallow. Following this rule, make as many different molecules as you can. A small marshmallow will go on the other end of the toothpick.

How many can you make? ___1___

Can you draw one?

❺-❻ Have the students compare the marshmallow molecules they made with and without rules. Help them find two shapes that are the same as each other and two that are different.

Help them record their observations.

III. What Did You Discover?

Read the questions with your students.

❶-❹ Have the students answer the questions in this section. Help them see that by following rules they were able to make fewer molecules than when they didn't follow rules. Explain to them that it is important for atoms to follow rules when making molecules in order for the same substances to be formed consistently from the same combinations of atoms.

IV. Why?

Read this section of the *Laboratory Notebook* with your students.

Discuss the concepts presented in this section of the *Laboratory Notebook*. Help the students understand that when people don't follow rules, there is no order — there is chaos. Help them see that the same thing would be true if atoms did not follow rules when forming molecules.

V. Just For Fun

In this section, students are given the choice of thinking about an existing game they play or making up a new one.

If students choose to think about an existing game, they are asked to think about rules that they follow when they play this game. Encourage them to think about why the rules are important to the game. Help them record whatever rules they think of.

Instead of thinking about an existing game, students can make up their own game and its rules. Allow the students to be creative and use their imagination. Help them record the rules of their new game.

Encouraging students to think about rules and their effects is the focus of this exercise. There are no right answers to this section.

Experiment 5

What Will Happen?

Materials Needed

- 4 or more clear plastic cups or glasses
- marking pen
- measuring cup
- measuring spoons
- the following food items:
    lemon juice - 180 ml (3/4 cup)
    vinegar - 180 ml (3/4 cup)
    milk - 180 ml (3/4 cup)
    baking soda - 90 ml (6 Tbsp.)
    water - 180 ml (3/4 cup)

Just For Fun section

- baking soda
- vinegar
- sugar

Or

- 2 or more food items chosen by student

Objectives

In this unit students will observe chemical reactions.

The objectives of this lesson are:

- To have students observe changes that occur in some substances when they undergo chemical reactions.

- To help students understand that not all substances will react chemically when mixed together.

Experiment

Setup—to do ahead of time:

❶ Label the plastic cups **A**, **B**, **C**, and **D**.

❷ Pour 60 ml (1/4 cup) of lemon juice into **Cup A**.

❸ Pour 60 ml (1/4 cup) of vinegar into **Cup B**.

❹ Pour 60 ml (1/4 cup) of milk into **Cup C**.

❺ Pour 60 ml (1/4 cup) of water into **Cup D** and add 30 ml (two tablespoons) of baking soda. Mix until the baking soda is completely dissolved.

(Each cup will be refilled twice with the same amount of the same liquid, or you can use a new cup each time, labeling it.)

Set the cups on a table. ***Do not tell the students what is in each cup.***

I. Think About It

Have the students examine the contents of each cup. They should observe the smell and color of the liquids and whether they are thick or thin. Have them write down what they observe, or record their answers for them.

II. Observe It

❶ **Cups A and B**

Ask the students what they think will happen if they add the contents of **Cup A** to **Cup B**. Have them write down their guess.

Next, have them pour the contents of **Cup A** into **Cup B**. Have them observe and record what happens.

Vinegar and lemon juice do not react, so they should not observe much happening.

Rinse the cups with water.

Refill **Cup A** with 60 ml (1/4 cup) of lemon juice and **Cup B** with 60 ml (1/4 cup) of vinegar.

❷ **Cups A and C**

Ask the students what they think will happen if they add the contents of **Cup A** to **Cup C**. Have them write down their answers.

Now have them pour the contents of **Cup A** into **Cup C**. Help them record what happens. The students will observe clumps forming in the milk as the lemon juice curdles the milk. This is a chemical reaction. The clumps are proteins in the milk that have been denatured (had their original properties changed) by the lemon juice. The clumps form a precipitate.

Rinse the cups with water.

Refill **Cup A** with 60 ml (1/4 cup) of lemon juice and **Cup C** with 60 ml (1/4 cup) of milk.

❸ **Cups A and D**

Ask the students what they think will happen if they add the contents of **Cup A** to **Cup D**. Have them write down their answers.

Have them pour the contents of **Cup A** into **Cup D**. Help them record what happens. They should observe a chemical reaction occurring between the lemon juice and the baking soda. The chemical reaction gives off bubbles which should be visible. If this does not happen, pour out the contents of **Cup D**. Make a new mixture of water and baking soda, and add twice as much baking soda. It doesn't matter whether or not all of the baking soda is dissolved. Then repeat the experiment.

Rinse the cups with water.

Refill **Cup D** with 60 ml (1/4 cup) of baking soda water.

❹ **Cups B and C**

Next, ask the students what they think will happen if they add the contents of **Cup B** to **Cup C**. Have them write down their answers.

Have them pour the contents of **Cup B** into **Cup C**. Help them record what happens. They should observe a chemical reaction similar to that of Step ❷. The vinegar should cause the milk to curdle.

Rinse the cups with water.

Refill **Cup B** with 60 ml (1/4 cup) of vinegar and **Cup C** with 60 ml (1/4 cup) of milk.

❺ Cups B and D

Next, ask the students what they think will happen if they add the contents of **Cup B** to **Cup D**. Have them write down their answers.

Have them pour the contents of **Cup B** into **Cup D**. Help them record what happens. They should observe a chemical reaction similar to that of Step **❸**. The vinegar and baking soda should react, and the mixture should give off bubbles.

Rinse the cups with water.

Refill **Cup D** with 60 ml (1/4 cup) of baking soda water.

❻ Cups C and D

Ask the students what they think will happen if they add the contents of **Cup C** to **Cup D**. Have them write down their answers.

Have them pour the contents of **Cup C** into **Cup D**. Help them record what happens. They should not observe any chemical reaction taking place. The baking soda will not make the milk curdle, nor will there be any visible bubbles or any other signs of a chemical reaction taking place.

Have the students pour out the contents of **Cup D** and clean up the experiment space.

Summary

Have the students summarize their results. The answers to questions **❶**-**❻** are provided below.

❶ Did lemon juice (**A**) react with vinegar (**B**)? _____ *no* _____

❷ Did lemon juice (**A**) react with milk (**C**)? _____ *yes* _____

❸ Did lemon juice (**A**) react with baking soda (**D**)? ___ *yes* _____

❹ Did vinegar (**B**) react with milk (**C**)? _____ *yes* _____

❺ Did vinegar (**B**) react with baking soda (**D**)? _____ *yes* _____

❻ Did milk (**C**) react with baking soda (**D**)? _____ *no* _____

III. What Did You Discover?

Help the students answer the questions in this section of the *Laboratory Notebook*. Their answers may vary depending on what they observed.

IV. Why?

Read this section of the *Laboratory Notebook* with your students.

Discuss the similarities and differences between the four liquids. Explain that vinegar and lemon juice are made of similar kinds of molecules, but milk and baking soda are not similar to vinegar and lemon juice. Although lemon juice and vinegar are not identical, they behave in similar ways because they are composed of similar kinds of molecules.

V. Just For Fun

Students can try mixing baking soda, sugar, and vinegar together in a cup. Let them experiment with how much of each substance to use. Adding sugar should result in a more impressive reaction. This experiment may be repeated using different amounts of the various substances.

Alternatively, students can look around the kitchen to find two other foods to try mixing together to see if they will react. This experiment may be repeated with other food combinations.

Have the students record their results in the space provided.

Experiment 6

Sour or Not Sour?

Materials Needed

- 12 (or more) clear plastic cups
- measuring cup
- measuring spoons
- marking pen
- one head of red cabbage
- knife
- cooking pot, large
- the following food items:
  distilled water, 1.25–1.75 liters
  (5–7 cups)
  white grape juice, 60 ml (1/4 cup)
  milk, 60 ml (1/4 cup)
  lemon juice, 60 ml (1/4 cup)
  grapefruit juice, 60 ml (1/4 cup)
  mineral water, 60 ml (1/4 cup)
  antacid tablets—3 extra-strength
  unflavored white Tums
  baking soda, 5 ml (1 teaspoon)
- other substances to test
  (see *Just For Fun* section)

Optional

- small plastic bag
- wooden mallet or other hard object
  for crushing antacid tablets

Objectives

In this experiment students will begin to explore the properties of acids and bases.

The objectives of this lesson are:

- To have students observe that acids and bases have different properties.
- To introduce the concept of indicators — in this experiment, red cabbage juice is used as an acid-base indicator to determine whether liquids are acids or bases.

Experiment

This experiment requires that students taste both acids and bases. It is relatively easy to find foods that are acidic but much more difficult to find foods that are basic. The only two safe products that we could find that are basic are baking soda and antacids. Most household cleaning products are basic, but these are not listed since they are not safe to taste.

Setup

NOTE: Do not use tap water for this experiment. Use only distilled water or you will not get the correct results.

To do 1 hour before the experiment

Chop or shred the head of red cabbage, and boil it in 1-1.5 liters (4-6 cups) of distilled water for 15 minutes. Remove the cabbage and allow the liquid to cool to room temperature.

Prepare liquids to be tested.

Dissolve the antacid tablets in distilled water. Add three extra-strength unflavored white Tums tablets to 60 milliliters (1/4 cup) of distilled water. Crushing the tablets may help in obtaining the color change of the cabbage juice indicator. To crush them, the tablets can be put in a plastic bag and hit with a hard object, such as a wooden mallet. If this mixture does not change the color, try adding more tablets. Other brands of antacids may or may not work.

To make the baking soda water, add 5 milliliters (1 teaspoon) of baking soda to 60 milliliters (1/4 cup) of distilled water.

Pour 60 milliliters (1/4 cup) of each liquid into a separate clear plastic cup, and using a marking pen, label each cup with the name of its contents.

I. Think About It

Read the text with your students.

Have the students make predictions about which liquids will taste sour and which will not taste sour. Help them mark their predictions in the proper column of the table in their *Laboratory Notebook*. Their answers may vary.

II. Observe It

Read this section of the *Laboratory Notebook* with your students.

❶ Have the students tear out the *Laboratory Notebook* pages that are labeled "SOUR" and "NOT SOUR" and place them on a table.

Have the students taste each liquid and indicate on the chart in the *Laboratory Notebook* whether it is sour or not sour. Help them try to distinguish between "sour" and "bitter." The mineral water and the baking soda water will taste "bad" but not sour. They are bitter or salty. The antacid water will taste sweet. Also, white grape juice may be sweet and not necessarily sour. Let the students decide whether they think it is sour or not sour. After recording each answer, have them place the cup of liquid on either the SOUR or NOT SOUR page, according to the taste.

Their answers may look as follows (answers may vary).

Liquid	Sour	Not Sour
white grape juice		X
milk		X
lemon juice	X	
grapefruit juice	X	
mineral water		X
antacid		X
distilled water		X
baking soda water		X

❷ Pour into a measuring cup the red cabbage juice that you made earlier and have the students observe the color of the cabbage juice. Then have them select one of the cups of liquid they tasted and observe the color of that liquid. Next, have them add 60 milliliters (1/4 cup) of red cabbage juice to the liquid.

❸ Ask them whether the color changes or stays the same. What they are looking for is ***the color change of the cabbage juice***. Its natural color is a deep red-purple. It will change to pink, green, or light purple when mixed with the other liquids.

Have the students return the cup to the SOUR or NOT SOUR page they took it from, and have them record their results.

❹ Have the students repeat Steps ❷-❸ for each of the liquids they tasted. Expected results are shown in the following chart:

Liquid	Color change? (yes or no)	What is the color?
white grape juice	*yes*	*pink*
milk	*no*	*purple*
lemon juice	*yes*	*pink*
grapefruit juice	*yes*	*pink*
mineral water	*yes/no*	*light purple*
antacid	*yes*	*green*
distilled water	*no*	*purple*
baking soda water	*yes*	*green*

III. What Did You Discover?

Have the students look at the cups that are on the SOUR and NOT SOUR pages. Ask them to observe the colors of the liquids and whether they see similarities or differences between those that are on the same page.

Help the students answer the questions in this section. Example answers follow.

(Answers may vary.)

❶ Which liquids were sour? *lemon juice and grapefruit juice*

❷ Which liquids were not sour? *milk, distilled water, mineral water*

❸ When you added the cabbage juice to the "sour" liquids, what color did the cabbage juice become? *pink*

❹ When you added the cabbage juice to the "not sour" liquids, what color did the cabbage juice become? *green or purple*

❺ Why do you think the "sour" liquids and "not sour" liquids turned the cabbage juice different colors? *They have different types of molecules.*

❻ If you added cabbage juice to a drink and it turned pink, do you think that drink would taste sour? *yes*

IV. Why?

Read the text with your students.

Discuss this section with the students. Have them think about why some of the liquids turned the red cabbage juice pink and some turned it green. Explain to them that the liquids that turned the cabbage juice pink are called acids, and the liquids that turned the cabbage juice green are called bases.

Explain that red cabbage juice is an indicator, which is anything that points out something to us. For example, a gas gauge in a car could be called an indicator—it tells the level of gas in the tank. The thermostat in a house could be called an indicator—it tells the temperature of the room.

In chemistry the term *indicator* refers to a chemical that tells you something about other chemicals. Red cabbage juice is an acid-base indicator, telling you whether the liquid is acidic or basic.

Explain that red cabbage juice will always turn pink in acids and will always turn green in bases unless there is something wrong with the indicator. Some liquids, such as milk and water, do not turn the indicator another color. Explain that these liquids are called *neutral*, and they are neither acids nor bases.

V. Just For Fun

Help the students find some other liquids to test with the red cabbage juice indicator. The students are **NOT TO TASTE** these liquids, so they can select some things like household cleaners that are not edible. They can also mix a powdered substance into distilled water and test the mixture.

Have the students decide whether the liquid is an acid, a base, or neutral, and help them record their observations.

Experiment 7

Pink and Green Together

Materials Needed

- 18 or more clear plastic cups
- measuring cup
- measuring spoons
- marking pen
- leftover red cabbage juice from Experiment 6 or one head of red cabbage
- the following food items, approx. 300 ml (1 1/4 cups) each:
  vinegar
  lemon juice
  mineral water
  distilled water (if you need to make red cabbage juice, you will need 1.5 liters more)
- baking soda, 25 ml (5 tsp.) or more
- antacid tablets, 5 or more (try Tums plain, white, extra strength)
- substances of students' choice to mix together

Objectives

In the last experiment students added red cabbage juice to several liquids to determine which were acids and which were bases. In this experiment students will continue their exploration of acids and bases.

The objectives of this lesson are for students to:

- Explore what happens when an acid and a base are mixed together.
- See that mixing an acid and a base can result in a neutral mixture, one that is neither an acid nor a base.

Experiment

If you have refrigerated red cabbage juice left from Experiment 3, use that. Otherwise follow the directions below.

If you do not have red cabbage juice from Experiment 6, do the following 1 hour before:

Chop or shred one head of red cabbage and boil it in approximately 1.5 liters (6 cups) of distilled water for 15 minutes. Remove the cabbage and allow the liquid to cool to room temperature.

NOTE: Do not use tap water. Use only distilled water or you will not get the correct results.

Setup

Put 60 ml (1/4 cup) of each of the following liquids into separate plastic cups and label the cups with a marking pen:

- vinegar
- lemon juice
- mineral water
- distilled water

Take two more plastic cups and put 60 ml (1/4 cup) of distilled water in each. Add 5 ml (1 tsp.) of baking soda to one cup and an antacid tablet to the other. (You may want to break or crush the tablet to help it dissolve faster.)

Alternatively, you can mix enough baking soda water and antacid water for the entire experiment. Use 300 ml (1 1/4 cups) distilled water to 25 ml (5 tsp.) baking soda and the same amount of distilled water with 5 or more antacid tablets. Then put 60 ml (1/4 cup) of each solution in a cup.

Label the cups.

I. Think About It

Read this section of the *Laboratory Notebook* with your students.

❶-❺ Have the students think about and answer the questions in this section of the *Laboratory Notebook*. Their answers will vary.

II. Observe It

Read this section of the *Laboratory Notebook* with your students.

❶ Place all of the cups on the table and have the students add 60 ml (1/4 cup) of cabbage juice to each cup. Have them observe the color of the liquid in each cup, and then have them record their results in the chart.

They should get the following:

Liquid	Pink	Green	Purple
distilled water			X
mineral water			X
lemon juice	X		
vinegar	X		
baking soda water		X	
antacid water		X	

❷ Have the students pour a green liquid into a cup containing a pink liquid and then a pink liquid into a cup containing a green liquid. This way they can observe that the result will be the same if an acid is added to a base or a base is added to an acid. Have them try all the combinations of pink and green liquids. Help the students fill more cups as needed.

Have them record their observations. Not all the empty squares will be filled in during this part of the experiment.

	antacid water	lemon juice	vinegar	mineral water	distilled water	baking soda water
antacid water	✕					
lemon juice	✕	✕				
vinegar	✕	✕	✕			
mineral water	✕	✕	✕	✕		
distilled water	✕	✕	✕	✕	✕	
baking soda water	✕	✕	✕	✕	✕	✕

❸ Next, have the students mix together the remaining liquids listed on the chart. Help them record any color changes that occur. For example, when lemon juice (pink) is added to mineral water (purple), the mineral water will turn pink. When mineral water (purple) is added to baking soda water (green), the color may change only slightly.

Encourage them to keep pouring the liquids back and forth to see what happens when mixtures are added to other mixtures. In the end, all of the liquids should turn purple. If some liquids are still green or pink, have the students pour them back and forth until every cup contains purple liquid.

Have them record anything they observe that they find interesting.

III. What Did You Discover?

Read this section of the *Laboratory Notebook* with your students.

❶-❹ Help the students answer the questions in this section. They should have seen some of the pink liquids turn green when green liquid was added and some of the green liquids turn pink when pink liquid was added.

IV. Why?

Read this section of the *Laboratory Notebook* with your students.

Have the students look at the chart they made and discuss the results with them. Explain that when they poured the liquids back and forth, the colors changed because the acids and bases were *reacting* with each other. Remind the students that in Chapter 6 they learned that a chemical reaction can be *indicated* by a color change. In this experiment the red cabbage juice indicator changed color as the acids and bases reacted with each other.

At the end of the experiment all of the liquids turn purple. Explain to the students that the acids and bases react with each other and cancel each other out, or *neutralize* each other. In the end there are no acids or bases left, only neutral liquids.

V. Just For Fun

Have the students look for some different liquids they can mix together and test with red cabbage juice indicator. Have them save these mixtures and then mix the mixtures together and see what happens. Do they change color? Have them observe whether they have an acid, a base, or a neutral mixture. They are not to taste these mixtures.

Make It Mix!

Materials Needed

- 15 or more clear plastic cups
- measuring cup
- measuring spoons
- spoon for mixing
- liquid soap
- marking pen
- the following food items [approx. 60 ml (1/4 cup) each]:
  water
  milk
  juice
  vegetable oil
  melted butter

Objectives

In this experiment students will observe mixtures.

The objectives of this lesson are to help students understand that:

- Liquids that are similar will mix.
- Liquids that are not similar will not mix.

Experiment

I. Think About It

Read this section of the *Laboratory Notebook* with your students.

Have the students think about whether each item in the top row is "like water" or "like oil" and then check the corresponding box below the item.

Have the students answer the questions in this section. They can refer to the chart they filled out. Their answers may vary and there are no "right" answers.

II. Observe It

Read this section of the *Laboratory Notebook* with your students.

Using clear plastic cups, help the students measure at least 60 ml (1/4 cup) of the following liquids into separate cups and label the cups with a marking pen:

- water
- milk
- juice
- oil
- butter

Using additional cups as test cups, have the students start mixing the liquids together by pouring about 15 ml (1 Tbsp.) of water into about 15 ml (1 Tbsp.) of milk. As the students make a mixture, have them use a marking pen to label the cup with the name of the mixture.

Have the students observe what happens when different liquids are mixed together. Then have them record their results in the chart in the *Laboratory Notebook*. Help them identify whether the two liquids mix or don't mix. When two liquids mix, the students won't be able to tell where one liquid starts and the other ends. When they don't mix, droplets of one liquid will be visible in the other.

Make sure the students do not confuse a color change with "mixing" or "not mixing." The liquids could change colors, but should be considered "mixed" only if there are no droplets visible.

It is not necessary to test every combination. At a minimum have the students test oil and water, oil and milk, and oil and butter.

Results should be as follows:

Results of Mixing Liquids					
	Water	**Milk**	**Juice**	**Oil**	**Butter**
Water		*mixed*	*mixed*	*not mixed (oil droplets visible)*	*not mixed (butter droplets visible)*
Milk			*mixed*	*slightly mixed*	*slightly mixed*
Juice				*not mixed (oil droplets visible)*	*not mixed (butter droplets visible)*
Oil					*mixed*
Butter					

Save the mixtures for the next part of the experiment.

To help students think about what they observed and to prepare them for the next part of the experiment, ask questions such as the following:

- *When you combined water, juice, and milk with each other do you think they mixed?*

- *If you were to add soap to any mixture of water, juice, or milk, do you think it would make any difference to how they mix?*

- *When you combined butter and oil with each other, did they mix?*

- *If you were to add soap to a combination of butter and oil, do you think it would make any difference to whether they mix?*

- *When would soap be needed in order to make two liquids mix together?*

Observe It With Soap

Using the mixtures from the first part of the experiment, have the students add about 2.5 ml (1/2 tsp.) of liquid soap to each mixture. The students should observe that soap doesn't change the liquids that already mix (e.g., water and juice), but does make the oil "mix" a little better into water and juice. It is not necessary to have them test all the mixtures, but have them add soap to a few of the oily mixtures and at least one of the mixtures of water-like liquids.

Their results will vary, but may look as follows:

Results of Adding Soap to Mixtures

	Water	Milk	Juice	Oil	Butter
Water		mixed	mixed	somewhat mixed	somewhat mixed
Milk			mixed	somewhat mixed	somewhat mixed
Juice				somewhat mixed	somewhat mixed
Oil					mixed
Butter					

III. What Did You Discover?

Read this section of the *Laboratory Notebook* with your students.

Help the students answer the questions in this section of the *Laboratory Notebook*. They should have observed that oil and butter do not mix with either water or juice. They also should have observed that oil mixes somewhat with milk and more with butter.

After adding soap, the students should have observed that oil mixes a little better with water and juice and much better with milk and butter.

IV. Why?

Read this section of the *Laboratory Notebook* with your students.

Have a discussion about the concepts presented in this section of the *Laboratory Notebook*. Explain to the students that "similar" liquids mix well, while liquids that are not "similar" do not mix well. Juice is similar to water because juice is mostly water, so juice and water mix well. Milk is a colloid, but will still mix well with water and juice, because milk is mostly water. (A colloid is a mixture that has very small droplets of molecules that do not actually mix well, but the droplets are so small it looks mixed. Colloids are often opaque). Oil and butter are similar because both oil and butter are fats. Oil and water are not similar, so oil will not mix well with either water or juice.

Explain the "rule" that similar liquids mix and dissimilar liquids do not mix.

Explain that soap is both a little bit like water and a little bit like oil, so soap mixes in both types of liquids. Because soap is like both water and oil, it "dissolves" oil in water. This is why soap works as a cleaner.

V. Just For Fun

Students are to create their own experiment to find out if certain liquids are "like water" or "like oil." Have them think of ways they could test the liquids. For example, they might choose water for the fifth liquid and stir each of the other liquids into the water to see if they mix. Allow them to try their idea even if you know it won't work. Experiments that don't work can provide valuable information for scientific research.

Substitutions can be made for items on the list. Students can look around the kitchen to see what's available. Have them think about whether they need to do anything different to test thick items like mayonnaise.

Have the students give their experiment a name and make a chart to record their observations. They can fill in the chart provided or create their own. Have them make notes about their idea for the experiment, how they performed the experiment, and how well they think their idea worked.

Make It Un-mix

Materials Needed

- several glasses or plastic cups
- measuring cup
- 3 bags (small paper or plastic)
- several small rocks (5-10)
- Legos (handful)
- sand (2 handfuls)
- sugar (handful)
- salt (2 handfuls)
- water
- food coloring, several colors
- 1-2 white coffee filters
- white paper, several sheets
- scissors
- several pencils
- tape

Objectives

In this experiment students will explore techniques used for separating various mixtures.

The objectives of this lesson are to have students:

- Gain a basic understanding of mixtures and the separation of mixtures.
- Explore different ways of separating mixtures of large, dissimilar items.
- Find ways to separate mixtures that have small, similar components.
- Experiment with a technique called chromatography that can be used to separate molecules from mixtures.

Experiment

I. Think About It

Read this section of the *Laboratory Notebook* with your students and discuss the questions with them. Help them think of things they might do to separate several different kinds of mixtures. Their answers may vary. Encourage them to think of different "tools," such as a sieve or flour sifter for separating mixtures. Also guide them to think of using water to dissolve part of a mixture, such as salt in the salt/sand mixture. There are no right answers to these questions.

II. Observe It

Read this section of the *Laboratory Notebook* with your students.

Have the students test one or more of their own ideas for separating each mixture. Even if you know their idea won't work, let them test it. Answers will vary—possible answers follow.

❶ Take a handful of rocks and a handful of Legos and mix them together on the table. Now try to un-mix them. Draw or describe what you did.

used hands and fingers to un-mix the rocks from the Legos

used a cardboard box with holes in it to un-mix the rocks from the Legos

(Answers may vary.)

Rocks and Legos

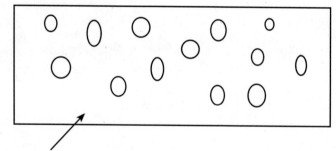

cardboard box with holes in it

❷ Take a handful of rocks and mix them with sand in a bag. Now un-mix the rocks and sand. Draw or describe what you did.

used a sieve to separate rocks and sand

used a hair dryer to blow away all of the sand

used cheesecloth to separate rocks and sand

(Answers may vary.)

Rocks and Sand

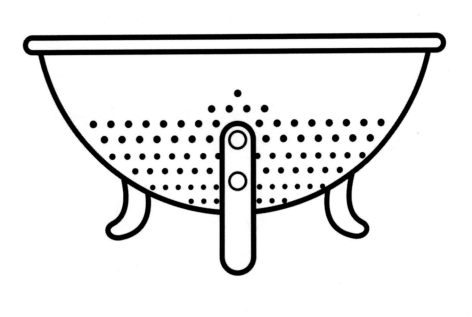

❸ Take a handful of sand and a handful of salt and mix them in a bag. Now un-mix them. Draw or describe what you did.

used water to dissolve the salt

(Answers may vary.)

Sand and Salt

salt water

sand

Paper Chromatography

The next steps in the experiment involve mixing different colors of food coloring and then separating the colors.

❹ Have the students place several drops of different colors of food coloring into 120 ml (1/2 cup) of water. The resulting color should be black or deep brown. Discuss possible ways to separate the colors.

❺ Next, discuss a method called *chromatography* that can be used to separate the colors from the water and from each other. Explain to the students that they can separate the colors by using a piece of coffee filter paper.

Help the students set up the chromatography sample. Have them cut the filter paper into long strips, tape one end of one strip to a pencil, and place the pencil across the glass containing the colored water, letting the paper strip dip into the water.

❻ They should observe the water immediately begin to migrate up the paper strip. They should detect the green food coloring migrating first, followed by the blue, then yellow, and finally red.

When they take the paper strip out of the water, have them lay it down on a piece of white paper. They should easily see the different colors. Have them record their results. When the paper strip is dry, it can be taped in the box in the *Laboratory Notebook*.

❼ Have them repeat the experiment with an "unknown." Without the students observing, add several drops of two or three colors into 120 ml (1/2 cup) of water in a glass. Give the glass to the students, and let them perform paper chromatography to determine which colors are in the water.

❽ Have the students prepare an "unknown" for the teacher, and let the teacher separate the colors. This is a lot of fun and can be repeated as many times as you wish.

❾ Have them record the results for Steps ❼ and ❽.

III. What Did You Discover?

Read this section of the *Laboratory Notebook* with your students.

Have the students answer the questions in this section of the *Laboratory Workbook*. Their answers will vary.

IV. Why?

Read this section of the *Laboratory Notebook* with your students.

Lead a discussion of the concepts covered in this section. Explain that there are many different ways to separate mixtures. Review the different ways the students discovered to separate the mixtures in their experiment.

Also discuss why some mixtures are easier to separate than others. Mixtures that have small components and mixtures that are made of similar items are harder to separate than mixtures with larger components and dissimilar items. Explain that scientists use a variety of tools and techniques to separate mixtures. The "trick" called chromatography is a technique frequently used by scientists to separate a variety of molecules. Explain that chromatography can be used to separate different kinds of molecules, such as proteins or DNA, and not just molecules that make color.

V. Just For Fun

Take a handful of salt and a handful of sugar and mix them in a bag. Now un-mix them. Draw or describe what you did.

(Answers may vary.)

Salt and Sugar

[This one is just for fun. Encourage the students to use their imagination.]

Used ants at a picnic

Since ants like sugar better than salt, the ants will carry off the sugar and leave the salt.

Experiment 10

Making Goo

Materials Needed

- Elmer's white glue, approx.
  30-60 ml (1/8-1/4 cup)
- liquid laundry starch, approx.
  30-60 ml (1/8-1/4 cup)*****
- measuring cup
- 2 plastic cups
- 30 metal paperclips
- *Just For Fun* section:
  non-toxic glue such as blue glue,
  clear glue, wood glue, glitter
  glue, or paste glue, approx.
  30-60 ml (1/8-1/4 cup)

  *If you are unable to find
  liquid laundry starch, you can
  use a mixture of equal parts
  cornstarch and borax mixed with
  enough water to dissolve them.
  Make about 30-60 ml (1/8-1/4
  cup) for this experiment.

Optional

- food coloring

Objectives

In this experiment students will explore the chemical reaction between Elmer's glue and liquid laundry starch.

The objectives of this lesson are to:

- Introduce the concept of polymers.
- Have students observe a chemical reaction that changes the properties of two polymers by the formation of cross-links.

Experiment

I. Think About It

Read this section of the *Laboratory Notebook* with your students.

Have the students think about the questions in this section of the *Laboratory Notebook* and encourage them to explore their ideas freely. Their answers may vary from those you expect. There are no right answers.

Guide inquiry with questions such as the following:

- *When you use glue, what does it feel like if you get it on your fingers?*

- *If you put your fingers in laundry starch (or cornstarch/borax mixed with water), what does it feel like?*

- *If you mixed glue and laundry starch together, do you think they would stay the same or change? Why or why not?*

- *What do you think the mixture would be like? Sticky? Slippery? Watery? Thick? What else might it be like?*

II. Observe It

Read this section of the *Laboratory Notebook* with your students.

In this experiment students will be adding liquid laundry starch or a cornstarch/borax/water mixture to Elmer's glue. Have the students measure 30-60 ml (⅛-¼ cup) of glue and pour it into a plastic cup. It is important not to put too much glue in the cup since students may need to add more laundry starch (or cornstarch/borax) than glue.

Have the students add 30-60 ml (⅛-¼ cup) laundry starch (or cornstarch/borax) to the glue. Nothing will happen until the students knead the glue and starch mixture. Have them add more laundry starch (or cornstarch/borax) if necessary. Encourage students to knead the mixture with their fingers. This is messy for teachers but delightful for most students. Both the glue and starch are nontoxic and can be easily cleaned from clothing and hands.

As students knead the mixture, help them think about what changes they may be observing. They should feel the glue become less sticky and more rubbery. They should be able to get the glue to roll into a ball or flatten in their hands like a pancake.

III. What Did You Discover?

Read this section of the *Laboratory Notebook* with your students.

Have the students answer the questions by describing what they actually observed. They should have noticed a significant change in the properties of the glue when it was kneaded with the starch.

IV. Why?

Read this section of the *Laboratory Notebook* with your students.

The glue and the laundry starch (or cornstarch/borax mixture) are both *polymers*, which are long chains made of hooked together molecules. When these two polymers are kneaded together, a chemical reaction occurs, changing the properties of the polymers. In this case, the starch makes the long chains of molecules in the glue hook to each other. This is called crosslinking because it makes cross-links between the polymers.

To illustrate this principle, gather 30 paperclips. Have the students make three chains with 10 paperclips each. Have the students lay them side by side on the table. Show the students that they can slide the chains of paperclips past each other. Explain that this is how the glue behaves without the laundry starch. Now take the paperclips and hook two or three from one chain to two or three from another chain. This is a cross-link. Next show the students that they can no longer easily slide the chains back and forth with respect to each other. This illustrates the changes that occur when the starch is kneaded into the glue.

Slides easily : no cross-links

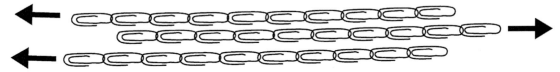

No sliding

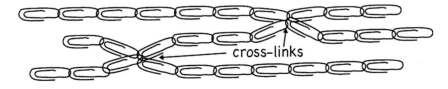

cross-links

V. Just For Fun

Students are to repeat the experiment using a different glue to see if this substitution changes the results. Help the students select a non-toxic glue.

If the students used liquid laundry starch in the original experiment, you can have them substitute the borax/cornstarch mixture for the laundry starch instead of using a different glue.

Students can add a few drops of food coloring to their mixture to see what it will look like.

Have them record their results.

Experiment 11

Salty or Sweet?

Materials Needed

- the following food items:
  marshmallows (2–3)
  ripe banana
  green banana
  pretzels or salty crackers,
    several
  raw potato
  cooked potato
  other food items
- blindfold

Objectives

In this experiment students will explore the concept that foods flavored by different molecules taste different.

The objectives of this lesson are to introduce students to the concepts that:

- Different molecules create different flavors in foods.
- The tongue has taste buds that sense different flavors.
- Long chains of carbohydrate molecules must be broken apart in order for the tongue to be able to taste the sugar molecules.

Experiment

To do 1 hour before:

Boil a raw potato until soft, then mash it and let it cool.

I. Think About It

Read this section of the *Laboratory Notebook* with your students.

Guide open inquiry with questions such as the following. There are no right answers to these questions.

- *Do you think you could taste anything if you did not have a tongue? Why or why not?*

- *Do you think you would enjoy eating food if everything tasted the same? Why or why not?*

- *How many different tastes do you think your tongue can detect?*

- *Do you think it makes a difference in the flavor of foods like potatoes and carrots if they are cooked or raw? Why or why not?*

- *If you chopped up raw vegetables, do you think they would taste different from when they are whole? Why or why not?*

- *What other indicators does your body have? (Guide the students to think of their five senses and go from there.)*

❶-❻ Ask the students to think about the questions in this section of the *Laboratory Notebook*. Help them record their answers. There are no right answers, and their answers may vary from those you would expect.

II. Observe It

Read this section of the *Laboratory Notebook* with your students.

Have the students tear out the pages labeled **SALTY**, **SWEET**, and **NEITHER** and spread them out on a table. Provide the food items to be tasted.

Have the students guess, *without tasting,* which foods will be salty, which will be sweet, and which will be neither. Have them place the foods on the corresponding pages.

Now take a blindfold and cover the students' eyes. Hand them one of the items from one of the pages, and ask them to guess if it is a sweet item, a salty item, or neither sweet nor salty. If they guess correctly, place the item back on the labeled paper. If their guess was incorrect, place the item off to the side.

When they finish tasting the items, remove the blindfold and have them see how many items they guessed correctly.

III. What Did You Discover?

Read this section of the *Laboratory Notebook* with your students.

❶-❻ Discuss the questions in this section. Help the students record their answers. Ask them how many foods they guessed correctly and how many they didn't guess correctly.

IV. Why?

Read this section of the *Laboratory Notebook* with your students.

Explain that the different flavors we taste in foods come from different molecules. The tongue is designed to detect these different molecules, causing the experience of different flavors. The taste buds on the tongue can tell salt molecules from sugar molecules.

Have a discussion about the fact that foods such as raw potatoes and green bananas contain long chains of sugar molecules called carbohydrates. Because the sugar molecules in carbohydrates are hooked together in long chains, taste buds cannot detect them. This is why raw potatoes and green bananas do not taste very sweet. Explain that when potatoes are cooked and bananas ripen, they become sweeter than when they are uncooked or not ripe. Cooking and ripening break apart carbohydrates (the long chains of sugar molecules) and then taste buds can detect the sugar.

V. Just For Fun

Have the students see if they get the same results if they repeat the experiment with someone else tasting the same foods. They might also like to find additional foods for comparison taste testing. Have them record their observations in the space provided.

Make It Rise!

Materials Needed

- flour, 2 liters (8 cups)
- 1 package active dry yeast, 7 grams (.25 oz.)
- lukewarm water, 240 ml (1 cup)
- cold water, 240 ml (1 cup)
- sugar, 30 ml (2 Tbsp.)
- vegetable oil
- salt, 5 ml (1 tsp.)
- soft butter, 120 ml (1/2 cup)
- double-acting baking powder, 15 ml (1 Tbsp.)
- milk, 360 ml (1 1/2 cups)
- measuring cups
- measuring spoons
- 4 mixing bowls
- mixing spoon
- floured bread board
- 2 bread pans or cookie sheets
- 2 cookie sheets
- marking pen
- refrigerator
- oven
- timer

Optional

- rolling pin
- biscuit cutter

Objectives

In this experiment students will observe how different temperatures affect the activity of enzymes in yeast.

The objectives of this lesson are:

- To introduce the concept of enzymes.
- To have students observe that enzymes must be within a certain temperature range in order to function properly.

Experiment

I. Think About It

Read this section of the *Laboratory Notebook* with your students.

Have the students answer the questions in this section of the *Laboratory Notebook*. Their answers will vary. Sample answers are given.

❶ List as many different types of molecules as you can.

acids, bases, salt, sugar, oils, water, enzymes

❷ What kinds of molecules make food salty?

salt molecules

❸ What kinds of molecules are glue and starch made of?

long chains (polymers)

❹ Do you think all the molecules in your body would work properly if your body got too hot? Why or why not?

(answers will vary)

❺ What kinds of molecules do you think make bread rise?

salt molecules? sugar molecules? acid molecules?

II. Observe It

Read this section of the *Laboratory Notebook* with your students.

Using the directions in the *Laboratory Notebook,* help the students make two rounds of bread dough. One dough will be made with warm water and placed in a warm place to rise—**Dough A.** The other dough will be made using cold water and placed in the refrigerator to rise—**Dough B.** The students should observe that **Dough A** rises and **Dough B** does not rise. Make sure that the warm water added to **Dough A** is not too hot. Hot water will kill the yeast.

III. What Did You Discover?

Read this section of the *Laboratory Notebook* with your students and have the students answer the questions.

Students should observe a significant difference between **Dough A** and **Dough B.** Help the students find words to describe what happened to the two doughs and what was different. Help them connect the fact that one dough was made with cold water and kept in a cold place, and the other dough was made with warm water and kept in a warm place. Point out to the students that temperature was the only difference between the two doughs. Help them see that this one change was what caused one dough to rise and the other dough not to rise.

IV. Why?

Read this section of the *Laboratory Notebook* with your students and discuss the concepts presented. Help the students understand that yeast is a living thing that contains very large protein molecules called *enzymes* that produce the gases needed for yeast to make bread rise. Yeast contains enzymes that convert sugar to carbon dioxide and alcohol. This is called fermentation. The carbon dioxide gas that is produced during the fermentation process is what makes the bread rise. The alcohol is burned off during the baking process.

Explain that there are many different kinds enzymes that perform a variety of tasks. Each enzyme is a large, complicated molecule that is shaped in a particular way and designed to perform a particular function. There are enzymes that cut molecules, enzymes that copy molecules, enzymes that glue molecules together, and enzymes that read other molecules.

Explain to the students that there are enzymes in their body that can only function within a narrow temperature range. The enzymes cannot function properly if the body temperature is either too hot or too cold.

V. Just For Fun

In this experiment students will observe what happens when biscuits are made with and without baking powder.

Baking Powder (or Not) Biscuits

Have the students think about what will happen when they make biscuits with and without baking powder. Have them record their ideas in the box provided.

Using the directions in the *Laboratory Notebook,* help the students make two biscuit dough mixtures. **Dough A** will be made with baking powder. **Dough B** will be made without baking powder. When they put the biscuits on cookie sheets, help them keep track of which cookie sheet contains **Dough A** and which has **Dough B.** The students should observe that, when they are baked, **Dough A** rises and **Dough B** does not rise.

Students should observe a significant difference between **Dough A** and **Dough B.** Have the students describe what happened to the two doughs and what was different. Point out to the students that adding baking powder or not adding it was the only difference between the two doughs. Help them see that this one change was what caused one dough to rise and the other dough not to rise.

Baking powder contains an acid and a base (often sodium bicarbonate, or baking soda) and when it gets wet, a chemical reaction occurs that releases bubbles of carbon dioxide. In recipes where baking soda is used, the recipe will call for the addition of an acid ingredient in liquid form (such as buttermilk or vinegar) to react with the baking soda.

Explain to the students that bread made with yeast and baking powder biscuits both rise due to release of carbon dioxide gas. However, the carbon dioxide in the bread is made by enzymes in yeast, a living thing, and the carbon dioxide from baking powder is made by a chemical reaction of non-living substances.

More REAL SCIENCE-4-KIDS Books
by Rebecca W. Keller, PhD

Building Blocks Series yearlong study program — each Student Textbook has accompanying Laboratory Notebook, Teacher's Manual, Lesson Plan, Study Notebook, Quizzes, and Graphics Package

Exploring Science Book K (Activity Book)
Exploring Science Book 1
Exploring Science Book 2
Exploring Science Book 3
Exploring Science Book 4
Exploring Science Book 5
Exploring Science Book 6
Exploring Science Book 7
Exploring Science Book 8

Focus On Series unit study program — each title has a Student Textbook with accompanying Laboratory Notebook, Teacher's Manual, Lesson Plan, Study Notebook, Quizzes, and Graphics Package

Focus On Elementary Chemistry
Focus On Elementary Biology
Focus On Elementary Physics
Focus On Elementary Geology
Focus On Elementary Astronomy

Focus On Middle School Chemistry
Focus On Middle School Biology
Focus On Middle School Physics
Focus On Middle School Geology
Focus On Middle School Astronomy

Focus On High School Chemistry

Super Simple Science Experiments

21 Super Simple Chemistry Experiments
21 Super Simple Biology Experiments
21 Super Simple Physics Experiments
21 Super Simple Geology Experiments
21 Super Simple Astronomy Experiments
101 Super Simple Science Experiments

Note: A few titles may still be in production.

Gravitas Publications Inc.

www.gravitaspublications.com
www.realscience4kids.com